CHICAGO PUBLIC LIBRARY
HAROLD WASHINGTON LIBRARY CENTER
R0003406962

D1613432

H.J.P. ARNOLD

Images from Space

The Camera in Orbit

Phaidon

Phaidon Press Limited, Littlegate House, St Ebbe's Street, Oxford

Published in the United States of America
by E. P. Dutton, New York

First Published 1979

ISBN 0 7148 2017 2
Library of Congress Catalog Card Number: 79–51327

Printed in Italy by Amilcare Pizzi, SpA, Milan

Preface

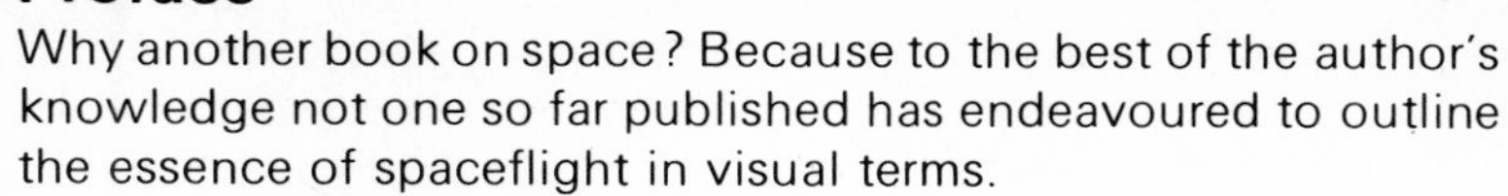

Why another book on space? Because to the best of the author's knowledge not one so far published has endeavoured to outline the essence of spaceflight in visual terms.

To select a mere sixty or so from the hundreds of thousands of pictures taken in space is a demanding, almost impossible and certainly daunting task: but it has been attempted here by someone who has had the privilege of being associated with the development of the techniques of imaging in space and of seeing many of the results; someone, moreover, who has a consuming desire to communicate the quintessence of these pictures to the man, woman and—of course—the youngster in the street. The images, obtained by means of photography, television or electronics, are complemented by the text and notes which situate the images in their context of place and time.

Despite the many Soviet achievements in space, very few mission photographs have been released and the quality of these has not been good. With the exception of one picture from the Apollo-Soyuz mission of 1975, therefore, none has been included here, though one hopes that, in the near future, the Soviet authorities will change their policy. The book is dedicated to the US National Aeronautics and Space Administration, its employees, and its industrial contractors, both in the United States and elsewhere. Whatever its failings and problems, and in the glare of unremitting publicity, NASA has laboured on the frontiers of space for two decades with conspicuous success, and has communicated the results freely to the world.

I Machines

With the dawning of the space age, the machinery of space fiction became fact. Rockets have achieved the velocities necessary to place spacecraft in orbit around the earth or on a trajectory into interplanetary space. The technology of the spacecraft developed quickly. Built to withstand the buffeting of the launch and the extremes of the space environment, unmanned spacecraft have taken man by proxy deep into the solar system, while man himself has journeyed into space in machines in which he has lived and worked for periods of days, weeks and even months and which have returned him safely to earth.

1. The launch of Apollo 15 in July 1971 as recorded by a protected automatic camera positioned at the 360 foot level of the mobile launcher. At this moment the five engines of the first stage of the Saturn V were building up to a total thrust of 7.7 million lbs. Above the silver-skinned service module and conical-shaped command module is the launch escape tower; this was equipped with a rocket motor which would have lifted the command module and the three astronauts to safety, had there been any serious problems during the launch.

2. More than six years after the Apollo 15 launch a Titan-Centaur rocket overcomes the earth's gravitational pull to send the unmanned spacecraft, Voyager 1, first to a rendezvous with Jupiter and then on to Saturn. On their voyage of exploration the two spacecraft in the Voyager project—like others before them in the programme of planetary exploration—are involved in what might be likened to a game of celestial billiards, where advantage is taken of Jupiter's gravity and the alignment of the planets beyond to send the vehicles on their way in a shorter time than would otherwise be possible. Moreover, the 'gravity assist' increase in velocity provided by Jupiter is available free of charge.

3. In the 1960s NASA sent unmanned spacecraft to the moon before risking sending men on the journey. The soft-landing Surveyors transmitted television pictures from the surface and sampled the soil—a vital task as some scientists believed strongly that there was a risk that the lunar module, carrying astronauts, might sink beneath a sea of dust. In April 1967 Surveyor III landed in the Ocean of Storms: two and a half years later the Apollo 12 lunar module made a safe precision landing within 600 feet of it. Parts of the vehicle were detached and brought back to earth for examination by specialists. In this Apollo 12 picture, Surveyor III is a solitary monument to man's technology. The trellis-like soil sampler can be seen on the right.

4. Project Gemini in 1965-6 perfected many of the techniques that were to be required during the Apollo lunar missions, in particular, the rendezvous and docking manœuvres. In June 1966 Tom Stafford and Eugene Cernan, the crew of Gemini IX, found that the shroud covering the target end of their docking vehicle had not separated and that it was still held by a restraining band, which made docking impossible. Here, eighty feet from Gemini IX, the target that became known as the 'angry alligator' appears to open its jaws menacingly to the sun's rays.

5. During lunar missions the Apollo command module was required to withdraw the carefully protected lunar module from the top of the third stage of the Saturn rocket. Apollo 7 in October 1968 did not go to the moon but it rehearsed certain parts of this manœuvre. In this photograph the round, white disc within the open panels of the third stage simulates the target on a lunar module. Below, in this striking image, we can see the brilliant blue envelope of the earth's atmosphere and, lower still, clouds over open sea. Beyond the vehicle there appears to be a magnificent vista of stars: these are in fact particles of expended fuel and other debris reflecting sunlight.

6. An astronaut's eye view of the Skylab space station as the last crew to occupy it left in February 1974 after eighty-four days in space. Since that time Skylab has remained unmanned and therefore a wasting asset—largely as a result of cuts in NASA's budgets. Sometime in the months ahead the station will re-enter the atmosphere and burn up; it is so large, however, that some of the debris will fall to earth. Three crews of astronauts travelled to Skylab in 1973/4 and conducted an intensive and highly successful series of experiments concerned with the remote sensing of earth resources, solar physics, man's ability to withstand prolonged exposure to zero gravity and the processing of industrial materials in space.

The station in this photograph provides a study in asymmetry for the solar array wing, which should have been extended on the right, was torn off during launch. At the top are the four solar arrays of the Apollo Telescope Mount structure from which the solar experiments were conducted.

7. The shape of the future. The age of the expendable rocket was magnificent but wasteful; the age of the space shuttle—due to begin in the next few months—will, it is hoped, make going into orbit a routine and less expensive operation, since it will be based on a re-usable space plane. The shuttle will launch like a rocket, fulfil its tasks in orbit and then return as a high-speed glider to earth, where it will be made ready for another mission within a matter of days. Tests to establish the behaviour of Enterprise, the first shuttle, during the critical approach and landing manœuvres were conducted in 1977 when it was dropped from a Boeing 747 carrier plane. The Enterprise is shown here during one of the test flights as it banks over California's Mojave Desert. The diagonal slope of the horizon heightens the drama of an already exciting picture.

8. Whatever the shapes of the future, it is the machines of the Apollo programme that stand out in the memory of the first years of space travel. The Apollo 9 mission in March 1969 saw the command and service modules (in which men would travel to and from the moon) and the lunar module (in which two of the astronauts would descend to the lunar surface) separate from one another for the first time, though in the comparative safety of earth orbit. In this view from the lunar module, the docking probe mechanism in the nose of the command module can be seen; the object jutting out below the spacecraft at the bottom left-hand side is an antenna for radio and television communications with Houston.

9. The lunar module needed no sleek, aerodynamic shape since it encountered no atmosphere. It was, in essence, a pair of highly sophisticated boxes set on four landing-struts, the lower one serving as a launching pad for the upper box when the programme of lunar surface activity was completed.

For over thirty-three hours in February 1971, the lunar module Antares served as base for the two Apollo 14 astronauts, Alan Shepard and Edgar Mitchell, during their time in the Fra Mauro region of the moon. The bright, white, circular flare—described by the crew as 'jewel-like'—at the top of the module was caused by the lunar morning sun shining into the camera lens. The colourful plastic cladding around the descent stage was designed to protect it from the extremes of heat and cold. Beyond the US flag is an umbrella-shaped communications antenna.

II Man

The role of man in spaceflight is a controversial one. He can be seen either as the most complex, versatile machine imaginable, or as an extremely limiting complication. However one may choose to define man's part in space travel, though, it seems likely that his explorations will continue.

In space, the human being is challenged: not only directly by the hostility of the environment—a challenge which has been met and largely overcome—but also on a deeper, more personal level, for the conditions which he encounters in a cramped spacecraft and the stresses which he undergoes are an extreme test of personality and of his ability to tolerate his fellow crew members. Prolonged training in conditions simulating those he will have to endure, assists greatly in preparing an astronaut for the time in space. But ultimately it comes down to his ability to face his own personality and that of his colleagues, outside the boundaries of normal experience and the familiar conditions which exist on earth. This will be the most serious problem astronauts will face when, eventually, they journey to the planets, and beyond.

10. When in February 1962 John Glenn made the first American orbital spaceflight, eight months had already passed since President John F. Kennedy had announced the goal of landing a man on the moon and returning him safely to earth again. At the time, the task seemed enormous, if not insuperable. But from relatively primitive beginnings and by a step-by-step procedure which ultimately overcame even a tragic disaster, the goal was achieved. From the one-man Mercury capsule which demonstrated that man could survive the space environment for many hours and be brought back safely, to the far greater sophistication and vital success of the two-man Gemini vehicle, and finally to Apollo itself, man worked constructively with man and as one with the machines upon which life in space depended.

This picture takes us back to the John Glenn mission of Friendship 7 which lasted a little over five hours: an automatic motion picture camera captured the dramatic scene as the space-suited astronaut monitored his instruments.

11. The first spacewalk by a NASA astronaut was performed by Edward H. White on 3 June 1965 during the Gemini IV mission. A Soviet cosmonaut had accomplished the world's first spacewalk ten weeks before. Here, White floats beside the spacecraft tethered to it by a twenty-five foot long umbilical cord. The chest-pack strapped to his harness received oxygen via the umbilical line from the spacecraft for both breathing and suit pressurization. The spacecraft's height above the clouds of well over one hundred miles results in little impression of movement though it was travelling at around 17,000 miles per hour. In White's right hand is a small self-manœuvring unit intended to give him control of his movements in space. On top of the unit is a 35 mm camera. White was one of three astronauts who died in a fire during an Apollo command module training exercise at Cape Canaveral in January 1967.

12. One of the most dramatic of in-cabin photographs ever: Gemini XI pilot, Richard Gordon, is shown fully spacesuited during an equipment jettison. The hatch is open and the blackness of space appears at the top right-hand side. The lighting is typical of the stark, contrasty conditions which exist in space where there is no atmosphere and often no other objects to scatter or reflect the sun's brilliant rays as on earth. The Gemini XI mission took place in September 1966 and was commanded by Charles ('Pete') Conrad, who took this picture. Gordon's spacewalk programme was interrupted by the kind of problems experienced on earlier Gemini flights—these included fogging of the helmet visor, overheating of the spacesuit and astronaut fatigue.

13. Edwin ('Buzz') Aldrin performed three spacewalks totalling five and a half hours during the following mission, Gemini XII, and demonstrated that the problems experienced on previous occasions could be largely overcome by the use of judicious rest periods and the provision of handrails, foot-restraints and supporting waist-tethers. During one of the spacewalks Aldrin used a specially designed camera (the back of which can be seen at the lower left-hand side) to take ultraviolet pictures and spectograms of stars and starfields. James Lovell commanded the Gemini XII flight which took place in November 1966 and which brought the highly successful, two-year programme to a conclusion.

14. The Apollo 7 mission was one of the less widely publicized flights, but one which, nonetheless, achieved several notable 'firsts': for example, it saw the first flight of the space module which had been drastically modified after the accident in 1967 in which three astronauts lost their lives. Indeed, it was the overall success of the eleven-day mission which led to the decision to send Apollo 8 in December 1968 on man's first journey to the moon.

The crew of Apollo 7 consisted of Mercury and Gemini veteran, Walter Schirra, command module pilot, Donn Eisele, and lunar module pilot (although there was no lunar module), Walter Cunningham. An unusual feature of the flight was the exchanges between an irascible Schirra and mission control in Houston: these came as a surprise to those who had always thought of astronauts as completely emotionless automatons. This high contrast and much underrated portrait, taken in orbit, shows the face of Walter Cunningham partially framed by communications cables, and, behind him, the window of the command module; his face shows signs of the strain and fatigue of a long mission in space.

15. An excellent impression of the limited space available in the Apollo command module is conveyed by this photograph of the Apollo 9 crew at work during a training exercise. The space available in the module for each crew member was roughly equal to that of two telephone kiosks, though the absence of gravity meant that the astronauts could move smoothly and without effort, so that they did not suffer seriously from feelings of claustrophobia. Here the astronauts lie on couches as during actual launch from the Kennedy Space Centre, with the commander, James McDivitt, in the traditional command position on the right. The astronauts are facing the main instrument panel.

16. The morning of 16 July 1969, and with launch a few hours away Neil Armstrong, commander of Apollo 11, 'suits up' in a special building at the Kennedy Space Centre. Beyond him are 'Buzz' Aldrin and Michael Collins. The complex Apollo spacesuit was three suits in one: a water-cooled underwear garment, a pressure suit which retained an oxygen atmosphere despite the vacuum of space outside, and a thirteen-layer protective garment designed to safeguard the astronaut against micrometeoroids and radiation. For a few hours before the launch the astronauts breathed pure oxygen from a portable case. This was vital since, had they breathed normal air, nitrogen, dissolved in the blood, would have caused the 'bends' when they entered the low pressure environment of the command module at the start of a mission.

This picture is a reminder that thousands of dedicated workers, technicians and managers laboured long and hard to allow this small, select band of astronauts to journey into space.

17. The Eagle has landed, and Neil Armstrong and Buzz Aldrin are the first human beings to disturb the dust which has lain on the lunar surface for aeons. Many of the pictures taken by Armstrong will always be considered as classics of lunar surface photography. Here, rim-lit by the sun to the left, Aldrin unfurls the aluminium sheet of the solar wind experiment: behind him is the lunar module. Footprints can be seen in the foreground, also the linear tracks which were formed by the cable of the surface television camera. The shapes of light on the left were caused by the sun shining almost directly into the camera lens.

18. An abstract massing of blacks and whites was recorded when Apollo 12 commander, Charles Conrad, backed out of the lunar module Intrepid to descend the ladder attached to one of the module's landing-struts; he thus became, in November 1969, the third man to walk on the moon. This photograph was taken by crew member Alan Bean, from within the lunar module. On Contrad's back is the portable life-support system (PLSS) pack, which supplied the Apollo spacesuit with the oxygen and coolants necessary to maintain life during work activity away from the spacecraft. The thin spear of light at the top is the radio antenna by which he communicated with Alan Bean and mission control back on earth.

19. A fine example of the range of tone and detail that can be recorded in lunar surface photographs when they are carefully exposed under optimum conditions, and of the myriad tasks facing the astronaut in a very limited time. Apollo 14 commander, Alan Shepard, is fixing an extension handle to a geological core tube for sampling the top layers of the lunar surface. The small transporter which was pulled or pushed by the astronauts during their geological traverses can scarcely be seen for equipment; geological tools and sample bags are attached everywhere. At the extreme left is the tall handle of a colour stereo close-up camera and just to one side of Shepard's right shoulder is the battery-powered movie camera on a staff.

20. Another Armstrong masterpiece from the Apollo 11 mission and perhaps the most famous picture ever taken on the moon's surface. Buzz Aldrin faces Armstrong's camera and in his convex visor can be seen his own shadow, together with the reflections of the lunar module, the US flag, the television camera and Armstrong himself taking the picture. Unfortunately, no original still photograph was taken of Neil Armstrong when on the lunar surface. Those that do exist are poor quality copies from the black and white television transmission or colour motion-picture frames.

21. The sun, unimpeded by any atmosphere, shines harshly across the flat surface of the Ocean of Storms as one of the Apollo 12 astronauts carries a pallet of scientific instruments to a spot well clear of the area that would be affected by the take-off of the lunar module. Behind him is the lunar module Intrepid and a dish-shaped communications antenna. The rays of light and patches of colour—lens aberrations caused by the direct and blinding sunlight—create a strange, unreal atmosphere.

22. Apollo 15 commander, David Scott, lifts equipment from the first, powered land-vehicle to be used on the moon: the lunar rover. In this photograph, taken by lunar module pilot, James Irwin, we see Hadley Rille, a lunar valley some 60 miles in length, up to 1,300 feet deep and one mile wide; the sun shines from the right, brilliantly illuminating the left side of the valley, but leaving the other side in deep, impenetrable shadow.

23. During the Skylab space station missions that took place between May 1973 and February 1974, astronauts conducted almost forty-two hours of spacewalks—'extravehicular activity' or 'EVA' in the jargon. The initial activity was devoted to unplanned repair work on the damaged station, which saved the entire mission: the planned work outside the station was largely concerned with experiments, including the changing of film magazines in the battery of cameras located in the Apollo Telescope Mount (ATM).

In this picture Skylab 3 scientist/astronaut, Owen Garriott, places an instrument for the collection of interplanetary dust on one of the large, solar array panels which powered the ATM. Beyond Garriott and the ATM support structure is the stark blackness of space. Men floating outside space vehicles has been a favourite scenario in space-fiction, and was superbly realized in, for example, the beautiful motion-picture, *2001: A Space Odyssey*. But reality, as depicted here, yields nothing in impact and mystery to fiction.

III Earth

When machines and then men first went into space people were astonished at the details that could be discerned from heights of hundreds of miles, and at the insights given by a macro view of the world, which spread out continents as though on a map. The frequency with which unmanned satellites could be made to orbit over any point on earth, or the accuracy with which they could be placed in an apparently stationary orbit over a point on the equator from which they could observe continually an entire hemisphere, gave a new dimension to our ability to observe transient events, such as the weather.

The existing technology of 'remote sensing' from aircraft, using cameras and other instruments, underwent rapid development as part of the space programme, and now the satellite as a means of providing information about the resources of earth, is steadily becoming more accepted in the same way as the weather satellite and the communications satellite. For some, however, it is the emotional quality of the view from space which is the most significant: the sadness of seeing the spaceship, earth, menaced not by the blackness or the vacuum of space, but by its own inhabitants.

24. Almost 25,000 miles out on the way to the moon, the Apollo 8 astronauts, the first human beings to leave earth orbit, turned their cameras back to their home planet. Much of the western hemisphere is visible, from the mouth of the St. Lawrence River and nearby Newfoundland, to Tierra del Fuego at the tip of South America. Central as well as South America are prominent and a small area of the bulge of West Africa shows at the top right-hand side where night encroaches on day.

James Lovell, who flew on the Apollo 8 mission and who narrowly escaped death in space during the Apollo 13 mission, said: 'The vast loneliness up here is awe-inspiring and it makes you realize just what you have back there on earth. The earth from here is a grand oasis in the . . . vastness of space.'

25. Not a crescent moon but a crescent earth. The Apollo 11 crew took this delightful picture of the planet aptly called by some 'The Blue Planet'. The immediate, overall impression one gains from all the deep-space views of earth in colour is that of blue oceans and white clouds: its beauty at a distance seems completely unspoilt by the political, racial and religious boundaries created by its inhabitants. A traveller from another world would have to come much closer before he saw signs of human presence.

26. The great emotional impact of the first views of earth from deep space was heightened by the use of colour film. But much of the photography carried out during manned missions was of a technical nature and this frequently demanded the use of black and white film which could, for example, yield greater resolution, i.e. record finer detail. Many of the images brought back from the moon by Apollo 8 were of this kind and here is one. From an altitude of some 70 miles above the lunar surface, the camera records an earth-rise. On the extreme left of the disc of earth are the major cloud systems of the Antarctic continent while to the right, where the disc touches the lunar horizon, is the land mass of West Africa. In an exchange with Houston, Apollo 8 commander, Frank Borman, commented that the usual phrase about there being a beautiful moon out must be modified: 'We were just saying that there's a beautiful earth out there.'

27. Many fine photographs of the earth from orbit were obtained during Project Gemini in 1965-6. Reproduced here in black and white, and yet still remaining one of the most superb, is the image taken by Charles Conrad and Richard Gordon from an altitude of almost 500 miles during the Gemini XI mission. Below, are the subcontinent of India and, on the right, Sri Lanka. The western coastal strip of India is free of cloud but there is a continuous offshore line of cumulus and cloud covers the Western Ghats mountain range. Adam's Bridge between India and Sri Lanka can be clearly distinguished. The Himalayas are visible on the horizon and the high altitude affords an excellent view of the Indian Ocean and the Bay of Bengal. The almost homely intrusion into the picture of the target rocket's antenna provides a contrast with the overwhelming size of the subcontinent stretching out to the end of the earth.

28. For those of us who have never experienced spaceflight an impression of this experience is perhaps best communicated by images containing both sharp surface-detail and an extensive sweep of the earth's horizon, set in the blackness of interplanetary space. In photographic form such images create a vital feeling of depth and involvement which, for all their beauty, is missing from deep space views.

This is a fine example of such an image from the Apollo 12 mission. Sweeping in an arc across the lower half of the surface are Baja California and Mexico on the right or north, with Central America on the left. The Yucatan Peninsula is left of centre, just above the bottom of the frame and to the left (i.e. south) of the solid sheet of cloud which etches the Gulf coast of Mexico and southern Texas. The top half of the surface, as seen here, is occupied by the great expanse of the Pacific Ocean.

The dark, roughly rectangular-shaped object above the clouds close to the centre of the image is not a UFO, but one of the panels that protected the lunar module during lift-off and which were jettisoned when the command and service modules separated from the third stage of the Saturn rocket.

29. The spearhead of the fast-developing technology of remote sensing of the earth from space is the Landsat satellite, three of which have been launched so far with a fourth to follow in 1981. The information from the Landsats is being applied increasingly to the pressing problems of resource management, prospecting, pollution detection, optimum agricultural production, and the speedy composition of maps, to name just a few. But to the unknowing eye, the Landsat infra-red colour pictures present a strange, chromatic mosaic, an unreal world where trees and crops are pink or red, urban areas blue-grey and the sea often black (see p. 33).

This, however, is not a standard Landsat picture. Several images have been mosaiced and enhanced at NASA's Goddard Space Flight Centre and the result converted to simulated natural colour by US Geological Survey specialists at Flagstaff, Arizona. The subject is worthy of such great attention. From 570 miles above, Landsat has recorded the twists and turns of the awe-inspiring Grand Canyon, etched by the Colorado River which, from Lees Ferry at the bottom right to the Shivwits Plateau at top centre, runs for well over 200 miles through the picture. We are presented with an image of great delicacy and yet power; one of the world's great natural features, perhaps 30 million years old, recorded by one of the latest techniques to stem from man's creative genius.

30. The strange beauty of clouds viewed from orbit. In this Skylab 3 image, towering cumulus and cumulo-nimbus clouds rise above a lower cloud deck and the setting sun, shining from the right, causes their shadows to be thrown onto the clouds below. Night approaches from the left.

Pictures taken during manned space missions as well as astronaut observations have provided valuable information to supplement the continuous meteorological data supplied by orbiting, unmanned satellites. Not infrequently the pictures have also captured scenes of great aesthetic beauty.

31. The earth, from the European Space Agency's Meteosat weather satellite almost 22,500 miles above the equator at 0° longitude. The thermal infra-red image was recorded at noon on a day in May 1978 when much of Europe was enjoying bright sunshine. The highest cloud tops are reproduced in white whilst lower clouds appear as varying shades of grey. Cloud girdles much of the middle of the globe in the region of the inter-tropical convergence zone; and, in the southern hemisphere, massive cloud plumes reveal the location of the 'roaring forties'.

32. A jewel in the Caribbean. A high-resolution infra-red picture from Skylab 4 isolates the red necklace of the Berry Islands in the Bahamas—and a twenty-five mile wide sand-shoal in their lee. The shallowness of the water over the white sand is evident, but, beyond, the sea floor plunges to extreme depths.

The oceans of the world are receiving fast-increasing attention from both manned and unmanned spacecraft, and rightly so, for they cover over 70 per cent of the globe's surface, are the mainspring of the world's weather machine, are relatively little-known, and could hold the future for life on earth.

33. From the plains of the Ganges to the roof of the world; from the Indian border in the south to the Siwalik Hills of Nepal and the snow-clad Himalayas at the top: all pass in this image beneath a Landsat. The ruggedness of the Lesser Himalayas which rise to heights of over 10,000 feet is emphasized in mid-picture by the low angle of the winter sun, but the great crests to the north dominate the eye. Everest itself is at the edge of the picture at the top right-hand side and most of the approach routes followed by successive teams of climbers can be discerned. River valleys appear mostly as light-blue traces and the large, grey area at upper left is the Nepalese capital, Kathmandu.

34. This dramatic picture of the city lights of north-west Europe was taken at 10 p.m. on 2 April 1974 by a weather satellite. In England, London, Birmingham, Liverpool and Manchester are the most extensive light clusters, while the lights of Amsterdam, Rotterdam and The Hague are clearly discernible across the North Sea and those of Antwerp and Brussels to the immediate south. The lights of the Ruhr form a boomerang shape and those of Berlin shine through a thin covering of cloud at the top right-hand side.

Fog shrouds the north-east coast of England and Scotland while out in the Atlantic, to the west of the British Isles, a cold front extends southwards, developing into the loosely swirling clouds of a filling low to the south of Ireland. France is almost entirely cloaked by the clouds of an occluded front.

35. An image as delicate and high-key as the previous one was powerful and crude. The antenna at the University of Dundee's Electronics Laboratory received this infra-red picture from the NOAA 5 weather satellite on the morning of 18 May 1977. The countries of Scandinavia are rarely so cloud-free. The mountains are still largely snow-covered but they are etched by the dendritic-like tracery of the valleys and fjords.

IV Worlds Beyond

Man has been in person to the moon. His machines have gone to all planets as far as the orbit of Jupiter and we will soon have the first images and scientific measurements from a spacecraft near Saturn. The sun, too, is a continuing subject of study from space. The past two decades have been a time of great excitement for those investigating the nature of the solar system and the universe beyond; a time when it has been a privilege to be alive. The research is conducted for two main reasons: firstly, since from the other planets we may discover valuable information to help explain the complex processes of earth, and secondly to answer the eternal questions of What? How? Why?

36. Comet Kohoutek was a disappointment for we had been led to expect a clearly visible and beautiful object in the sky. It did exist, however, and was photographed extensively by the crew of Skylab 4, which was in orbit at that time. This is a colour version from an original black and white image exposed in a far ultra-violet electronographic camera, on special film, at wavelengths not possible on earth. Four density levels have been separated and each given a different colour to produce an abstract image. The original was exposed on Christmas Day 1973 when the comet's tail was some three million miles in length.

37. Although on 6 January 1979 when this picture was taken Voyager 1 was still almost thirty-six million miles from Jupiter, its narrow-angle, television-type camera was already able to discern more detail than the most powerful earth-bound telescopes, and the quality was such that it seemed certain that within a short time the most detailed images which had been obtained so far by the two Pioneer spacecraft as they flew by the solar system's largest planet in 1973 and 1974, would be surpassed. The complexity of Jupiter's atmosphere is the main feature of this image, which was built up from a number of pictures taken through different filters, the pictures being radioed back as 'bits' of data to NASA's large antennas and then reconstructed by computers. The famous Great Red Spot can just be made out at the extreme right 'limb' or edge of the planet. The smallest object resolved in this view is about 600 miles across: as was expected, at Voyager 1's closest approach early in 1979 details as small as a few miles across were recorded.

38. Like the Surveyors, Lunar Orbiters were elements in the programme of unmanned lunar exploration conducted by the US in the 1960s to secure the information needed before a manned landing on the moon could take place. This was dubbed the 'picture of the century' when it was received from Lunar Orbiter 2 in November 1966. It shows part of the massive crater, Copernicus. On the horizon is the Gay-Lussac area of the moon's Carpathian Mountains. It is about 150 miles from the horizon to the base of the picture and the distance across is some seventeen miles. The lines across the photograph result from the method by which the picture was transmitted to earth and reconstructed.

39. The last three Apollo missions carried large-format, high-resolution cameras which enabled the surface of the moon to be photographed with a precision never before attempted. To those used to the best that earth-bound telescopes can produce through the distortions of our atmosphere, results like this Apollo 16 image are a revelation. Craters of very small size can be seen, and this is particularly valuable as much of the image is of the far side of the moon, which never faces earth. Mare Crisium, which is at the far right of the moon as we see it, is the roughly circular shape at the extreme top left. The dark area below and to the right of Crisium is Mare Marginis and below again and to the left of that is Mare Smythii. The virtual absence of the dark mare material on the far side of the moon is striking.

40. Early in August 1976, and still over a quarter of a million miles away from Mars, a television camera aboard the Viking Orbiter 2 spacecraft took a colour image of dawn breaking over the Red Planet. The picture was subsequently improved by computer techniques and this is the superb result. In the north, trailing cloud plumes, is the giant volcano, Ascreaus Mons; in the centre is the 3,000 mile long rift canyon, Valles Marineris, and at the foot of the crescent is Argyre Basin, covered with the frosts of early morning.

41. Viking Lander 1 had been on the surface of Mars in the Plain of Chryse for over six months when this 110° colour panorama was taken in February 1977. Its meteorology boom extends upwards on the right and to one side are the trenches, up to one foot deep, dug by a telescopic arm which placed specimens of Martian soil in the spacecraft's miniaturized laboratory chambers. The rock in the middle distance is about 7 feet long and 26 feet from the Lander.

The apparent warmth, indeed heat of the scene should not mislead: the typical range of temperatures during the Lander's first weeks on the surface was from a low of −125 °F to a *high* of −21 °F.

42. In February 1974 the unmanned Mariner 10 spacecraft flew by Venus using the planet's gravity to bend its trajectory and decrease its speed so that it could cross the orbit of Mercury, the principal target of the mission, on 29 March and (in a fine display of interplanetary thrift) on 21 September, and yet again on 16 March 1975.

Mercury, the smallest planet in the solar system and the closest to the sun, is difficult to study from earth. Mariner

10's instruments revealed a highly-cratered surface with extensive areas strikingly similar to lunar highlands and maria, though there were also some important differences. To the casual glance this picture might seem to show the moon, but it is Mercury's south pole viewed from a distance of around 50,000 miles. The pole is located inside the 110 mile-wide crater to the left of centre; the crater floor is in shadow and its far rim, which is illuminated by the sun, appears to float in space.

43. Mariner 10 was the first space vehicle to transmit to earth detailed ultra-violet images of the clouds of Venus. A little under five years later, American spacecraft returned to the planet—a battery of probes descending to the surface and sounding the atmosphere until the moment they were destroyed, while another vehicle went into orbit to begin a lengthy examination of Venus. This picture, taken by the Pioneer Venus orbiter on 14 January 1979 from an altitude of 40,000 miles, shows approximately three-quarters of the planet illuminated by the sun, and reveals considerable detail of the enveloping clouds which extend from an altitude of approximately 42 miles down to a little over 30 miles. The dark markings are thought to be caused by the absorption of ultra-violet light by sulphur compounds.

Earlier called the earth's 'sister planet', Venus is in reality a hell planet—with temperatures of 850 °F at its surface and an atmosphere largely composed of carbon dioxide which is over ninety times as dense as on earth. On the surface, illumination is a lurid red with visibility little more than one and a half miles and with landmarks strongly distorted by refraction.

V Reality is Fantasy

Photography has played a variety of roles in the American space programme. It is a tool of high technology, but through it astronauts have sought also to record and communicate the majesty of the unfolding view. Such records can never be complete for they are one stage removed from the profound, personal and emotional experience of seeing for oneself. But many of the images which have been returned to earth have been excellent and have undoubtedly enriched our awareness of the universe. Sometimes the obtaining of such photographs has resulted from the imaginative eye of the astronaut; sometimes, it has resulted from chance—for example, reflections on windows—which only serves to demonstrate that serendipity is not confined solely to earth. Furthermore, impact and appeal are found also in some of the images taken by unmanned spacecraft.

Here are just a few of the images which best suggest the true nature of the space experience and, as such, could claim to be the visual poetry of the space era.

44. A full moon rises to greet the Skylab space station but refraction through the atmosphere causes an intriguing distortion. Below are clouds, above, the blackness of space, and, stretching away on either side of the moon, a cross-section of the beautiful, luminous blue of the earth's upper atmosphere.

45. From the time of the early Mercury flights, photographs have been taken of the earth's atmosphere as the sun shines through it, close to the horizon. Differential scattering of the sun's rays at different altitudes results in a broadly three-layer effect: blue at the top of the atmosphere, then white, with orange/red at the lowest level. This is one of the best examples of *twilight layering*, taken by the Skylab 3 astronauts, Alan Bean, Owen Garriott and Jack Lousma from an altitude of 270 miles.

46. Earth comes between the sun and the Apollo 12 astronauts returning from their lunar landing in the Ocean of Storms, creating a solar eclipse for their eyes alone.

47. The object looks like a work of imagination, a strange world appearing on the video screens of the USS Enterprise. But it is the reality of space: a high-resolution image of Phobos, one of the two moons of Mars, secured in 1976 by Viking Orbiter 2. The small, roughly oval-shaped (thirteen miles by twelve miles) satellite is a dead world, heavily cratered, but also bearing previously undetected striations which provide the scientists with yet another puzzle.

48. An image produced electronically, but with the overtones of an Impressionist painting. The ascent stage of the Apollo 15 lunar module lifts off from the plains at Hadley and sends multi-coloured debris flying in all directions. This was the first time ever that this part of a mission had been witnessed from earth. It was recorded by a remotely controlled television camera mounted on the lunar rover parked 300 feet away.

49. This is a false-colour version of a far ultra-violet picture of earth. The original was taken from the lunar surface by the Apollo 16 astronauts in April 1972 using a special camera, very similar to the one which obtained the images of Comet Kohoutek some time later from Skylab.

On the sunlit (red) side of earth can be seen ultra-violet day airglow which results from the absorption of far ultra-violet radiation from the sun by the atoms and molecules in the upper atmosphere. On the dark, right-hand side, however, additional emission phenomena can be seen: near both polar regions are auroras, while there are also two bands of airglow symmetrically placed on opposite sides of the terrestrial magnetic equator.

50. The steel and aluminium architecture of the moonflight era. Illuminated by searchlights, a Saturn V rocket is in position at Kennedy Space Centre's launch complex 39, the starting point for all America's manned lunar flights. Even the immense 363 foot high Saturn vehicle is dwarfed by the 446 foot tall mobile launcher which served both as an assembly platform during the rocket's construction and as a launch platform. A lattice-work of lights illuminates the different levels of the tower and the swing-arms which connect with the Saturn rocket.

51. In orbit around the moon, the Apollo 10 lunar module, brilliantly lit by the sun against the blackness of space, appears to be surrounded by mysterious lunar clouds and shapes. They are, in fact, internal reflections on the command module window through which the picture was taken by astronaut, John Young.

52. A golden dawn: morning sunlight is reflected by the Gulf of Mexico and, on the horizon, the Atlantic Ocean. The Florida peninsula appears as a largely dark silhouette in the middle of the photograph although reflections reveal graphically the considerable extent of surface water in the state. 'Streets' of cumulus cloud are prominent features over the Atlantic and stratus cloud extends over the Gulf to the right. The picture was taken by the Apollo 7 astronauts from an altitude of 140 miles.

Some may consider it wrong to analyse an image of this kind in terms of the earth sciences for it may be regarded also as a profound, spiritual statement.

53. Fifty feet away from Gemini XI an Agena target vehicle appears motionless, as a tether—used to connect the two craft during station-keeping exercises—floats free. The accidental exposure of the film to heat and light during the mission affected the chemical make-up of its colour layers, resulting in this impression of mysterious red mists in space.

54. 20 August 1976. Viking 1 records sunset over Chryse Planitia on Mars. One of the Lander spacecraft's cameras began scanning the scene from the left about four minutes after the sun had dropped below the horizon, continuing for ten minutes and covering 120°. The beautiful haloes in the sky do not exist in reality: they result from the manner in which the electronic camera system aboard the Viking Landers separated different levels of light intensity.

55. An artificial solar eclipse experiment was carried out during the joint US/Soviet space mission in July 1975. The major objective was to photograph the extended region of the solar corona against the blackness of space, from the Soyuz spacecraft, while the disc of the sun was occulted by the Apollo spacecraft as it manœuvred, first away from and then toward, the Russian vehicle, in what was literally a close encounter.

56. Skylab's Apollo Telescope Mount was used to obtain approximately 200,000 images of the sun and the experiments carried out have had a major effect on our knowledge of the sun and its processes. At the same time as advancing knowledge, however, many of the images, subsequently modified by computer or in the photographic laboratory, have provided beauty, mystery, splendour and even the exotic for the non-specialist eye to behold. This picture was taken during a period when active regions were rotating across the sun's disc, producing many flares. To the researcher it is a colour-coded guide to the active regions at a precise period in time; to others its meaning and appeal will be limited only by the boundaries of the imagination.

57. In a beautifully conceived picture, an Apollo 9 lunar module 'thrust-chamber assembly cluster' (the small propulsion units which were situated at each of the vehicle's corners) assumes the shape of a spaceship in its own right. Colours, created by bright light falling on the surfaces of the camera lens, add to the atmosphere of fantasy.

VI Vision of the Future

The space artist has an honoured role in the history of spaceflight: he goes where cameras follow. For years spaceflight existed mainly in the imagination of the artist and it was he who provided an inspiration for the engineers and other specialists who made spaceflight a reality. In projects which are in existence or planned, the artist is a highly talented, technical illustrator. In depicting the vistas of far places, now or in decades and centuries to come, the imagination, nurtured but not restricted by contemporary knowledge, is given free rein.

58. An instrument-laden Voyager spacecraft flies by Saturn. Two Voyagers are on the way to the solar system's second largest planet, with its intriguing and beautiful rings. The first is scheduled to reach it in November 1980 and the second in August 1981. Voyager 1 will pass within 2,500 miles of the surface of Saturn's (and the solar system's) largest moon, Titan—a favourite subject of space artists—while Voyager 2 can be directed on by its controllers to a January 1986 encounter with Uranus.

59. In a world consuming more and more energy and yet with rapidly diminishing stocks of hydrocarbon fuels, alternative sources of energy are receiving intensive study. One concept is that of solar power satellites, orbiting some 23,000 miles away from earth, where they will be in sunlight for 99 per cent of the time and where they will be able to tap the immense energy streaming from the sun. Electricity generated in space will be converted to microwave form for transmission to earth, where it will be re-converted into electricity.

Many of the solar power satellite proposals involve the use of arrays of photovoltaic cells to produce electricity. Shown here is another concept: four parabolic dishes (each about three and a half miles across) directing sunlight onto solar furnaces. In the furnaces gases are heated to extremely high temperatures by the concentrated sunlight and drive generators which, in turn, produce electricity for transmission to earth. In this impression, the almost fifteen-mile long, four-segment array has been completed and is on station. Around the four, domed furnaces are panels radiating unwanted heat into space and, at each end of the array, a transmitting dish for beaming microwaves to earth. Such a satellite would weigh 110,000 tons and generate 10,000 megawatts.

60. The space shuttle will have an enormous cargo-hold measuring sixty feet by fifteen feet, which will be used partly for transporting satellites into orbit and partly as a platform for experiments and Spacelab, the laboratory being constructed by the European Space Agency. In this impression of a typical scene of the 1980s, shuttle crew-members are working on a large satellite before it is activated and placed in orbit.

61. Our atmosphere both protects life on earth and seriously impedes scientific study of the universe beyond. In orbit above the atmosphere, new worlds—both literally and metaphorically—will be opened to our gaze. Scheduled for insertion into orbit from the shuttle in 1983 is the space telescope, a NASA/European Space Agency project, which will enable astronomers and other scientists to see that which was never before possible (objects ten times smaller than those detected with surface-based telescopes); to probe distances ten times further from earth, and to detect objects fifty times fainter. The telescope will certainly play a dominant role in astronomy in the next few decades and at present it is impossible even to imagine the revolution it may lead to in our knowledge and awareness of the universe.

62. The most imaginative yet nonetheless practical space concept advanced in recent years is that associated with the name of Professor Gerard K. O'Neill of Princeton University and an ever-widening group of specialists and researchers: namely, that of colonies on the 'high frontier' of deep earth orbit, containing at first hundreds, then thousands, tens of thousands and eventually hundreds of thousands of people. Although involving a huge investment, the scheme has been costed in realistic terms: the mining of first the moon and then the asteroids for much of the material with which to build and run the habitats; the use of an electro-magnetic 'mass-driver' to deliver the raw materials into orbit; and the sound development of the colonies' economies by concentrating on products for which they are best suited and in which the earth could not rival them—for example, the construction of solar power satellites.

This artist's impression shows one of the advanced, colony-structures. Each cylinder in the pair is nineteen miles long and four miles in diameter, rotating on its axis to create earth-like gravity. The cylindrical section of the colony is the living area which could be built to resemble a terrestrial landscape. The tea-cup-shaped containers ringing the cylinder are agricultural stations and the installations protruding as hubs are manufacturing and power stations. Massive, rectangular mirrors on the sides of the cylinders would direct sunlight into the interiors, regulating both the seasons and the day/night cycle.

63. Many scientists now take it for granted that there are intelligent life-forms elsewhere in our galaxy or in the galaxies beyond. The vast distances involved make direct contact improbable, but planets inhabited by intelligent life would, by definition, be sources of radio signals travelling outwards at the speed of light. Various strategies have been proposed in what has come to be called SETI (the search for extraterrestrial intelligence), one of which is Project Cyclops. Cyclops would comprise, ultimately, between 1,000 and 2,500 radio antennas in a huge array of about ten miles in diameter, controlled and steered by a computer complex. This view shows a segment of the array with the central control and processing building on the right.

64. In some parts of the universe galaxies are clustered close together, so that the chances of a collision are around once in 15,000 million years: once in approximately the lifetime of a galaxy. Even where actual collisions do not occur galaxies can pass close to one another and the observation of such an event provides astronomers with valuable information about their structures. Here, in the sky of an imaginary, earth-like world painted by Don Dixon, two galaxies drift past one another drawing streams of particles and gas across the intergalactic space. Recent experiments with computer models have suggested that such *tidal distortions* may be responsible for producing spiral structures in galaxies.

Acknowledgements

In almost fifteen years of researching and writing on space, the pictures of space and how they were made, I owe a great debt to many people, some of whom I have space enough to name below:
John Brinkmann, Richard Underwood, Verl Wilmarth, John McLeaish, Douglas Ward and Sandra McGee (NASA: Lyndon B. Johnson Space Centre, Houston)
Les Gaver (NASA Headquarters: Washington)
Ed Harrison (NASA: Kennedy Space Centre, Florida)
Frank Bristow and Don Bane (NASA: Jet Propulsion Laboratory, California)
Donald Witten (NASA: Goddard SFC, Maryland)
Joseph Jones (NASA: Marshall SFC, Alabama)
Sandee Henry (NASA: Ames Research Centre, California)
James Sullivan (Naval Research Laboratory, Washington)
William Dunn (US Embassy, London)
Librarians of the Space Science and Engineering Centre (University of Wisconsin at Madison)

Sources of Illustrations

NASA: 1, 2, 3, 4, 5, 6, 8, 9, 10, 11, 12, 13, 14, 15, 16, 17, 18, 19, 20, 21, 22, 23, 24, 25, 26, 27, 28, 30, 32, 37, 38, 40, 42, 43, 44, 45, 46, 47, 48, 50, 51, 52, 53, 55, 56, 57, 58, 60, 62, 63.

Rockwell International/NASA: 7
NASA/US Geological Survey: 29
Meteosat/ESA: 31
General Electric Company: 33
US Air Force Weather Service—DMSP Data: 34
University of Dundee, Electronics Laboratory: 35
US Naval Research Laboratory: 36, 49
Fairchild Space and Defence Systems: 39
Martin Marietta Aerospace/NASA: 41, 54
Boeing Aerospace: 59
Lockheed Missiles and Space: 61
Don Dixon: 64

UR COMPLEX 41
NASA TITAN/CENTAUR LAUNCHES

UNITED
STATES

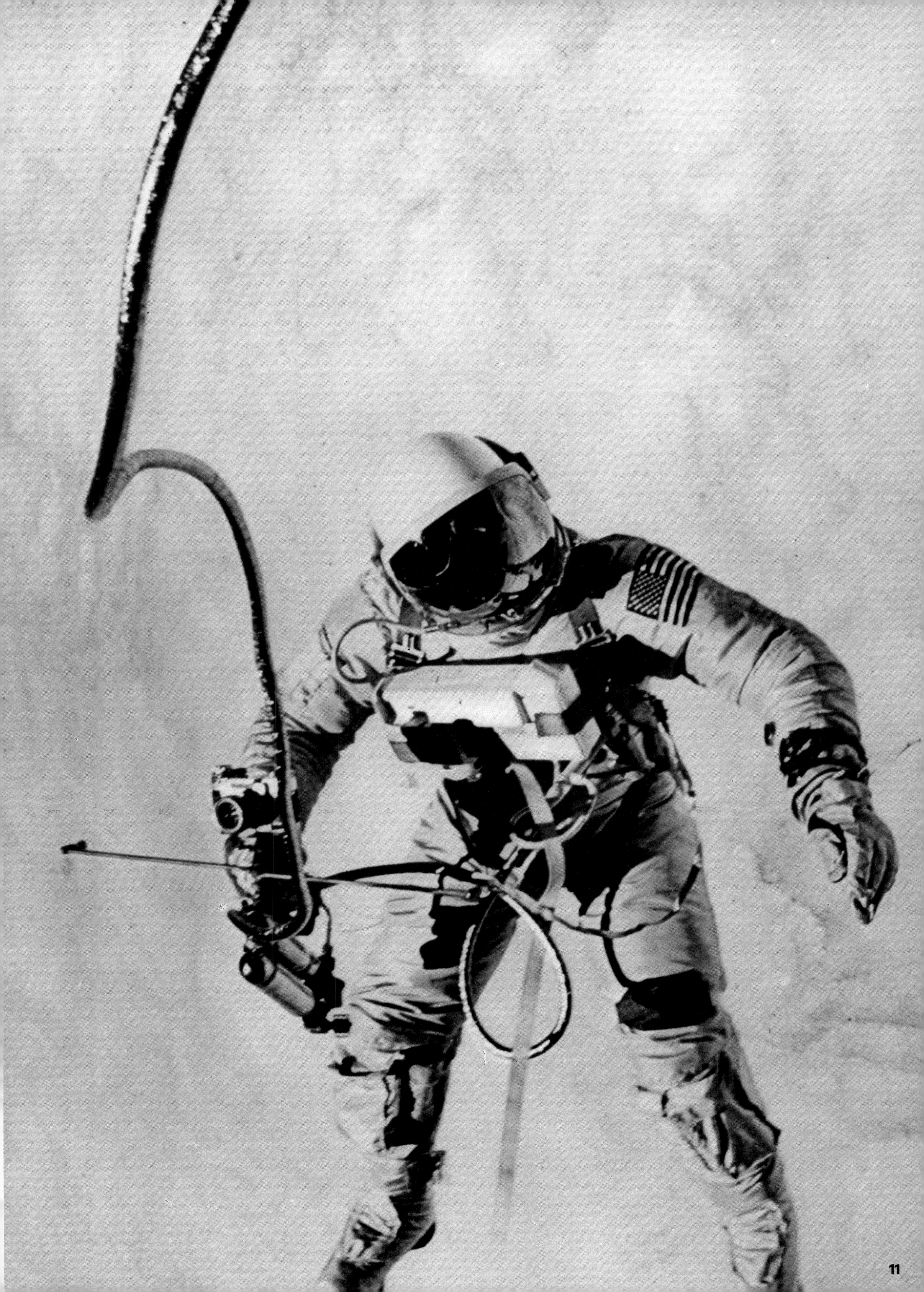

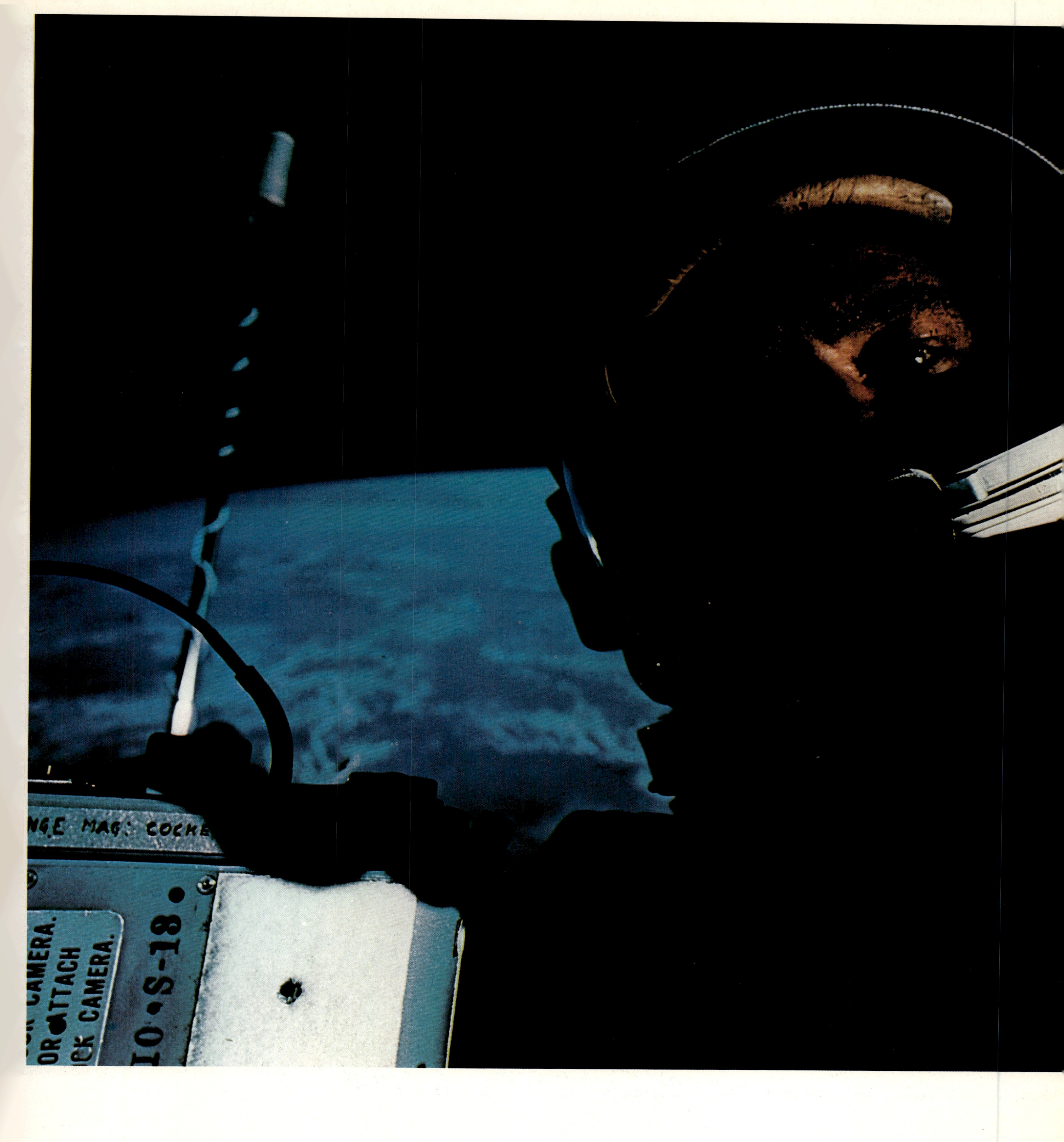
NGE MAG: COCKE
CAMERA.
OR ATTACH
CK CAMERA.
IO S-18

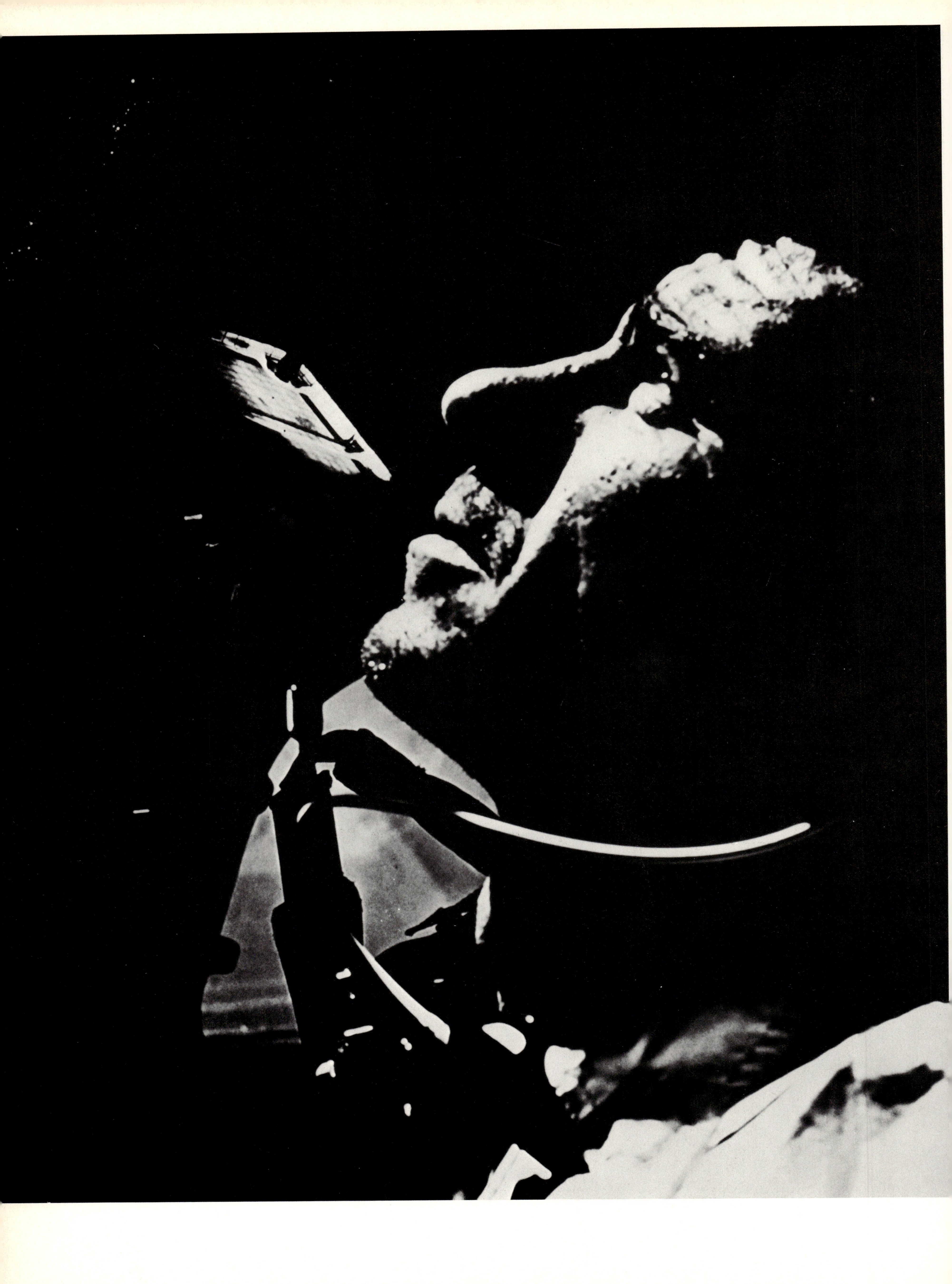

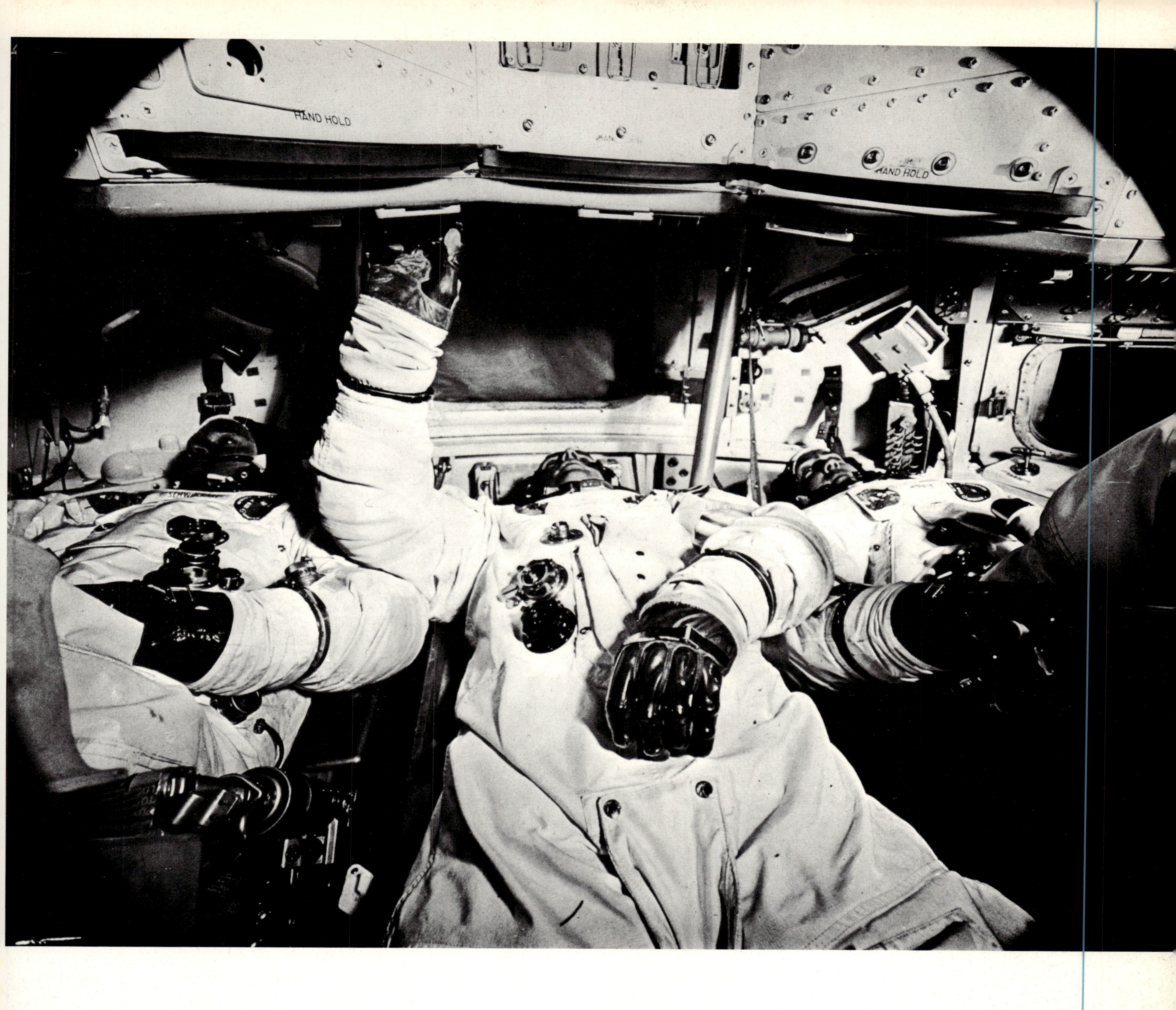
HAND HOLD
HAND HOLD